FLORA OF TROPICAL EAST AFRICA

OPILIACEAE

G. Ll. Lucas

Trees, shrubs, or woody climbers. Leaves alternate, entire, simple, exstipulate. Inflorescence racemose or umbellate. Flowers mainly bisexual but some dioecious (e.g. in *Agonandra*), but not in Africa, regular. Sepals 4–5, free, valvate, or sometimes partially united. Petals 4–5, free, valvate or sometimes partially united. Stamens opposite and equal in number to the petals, free or partially united to them; anthers dithecous, dehiscing longitudinally. Disk entire or bearing free glands alternating with the stamens. Ovary superior to semi-inferior, unilocular; ovule solitary, pendulous from the apex of a central placenta; stigma sessile or borne on a slender style. Fruit a drupe, often fleshy. Seeds having copious endosperm and a relatively small embryo.

A small family, formerly included in the Olacaceae, limited to the tropics, with the majority of representatives in Asia and Africa, and two genera in East Africa.

NOTE. The branched system of lignified cells in the leaves and the presence of cystoliths in all genera separate this group from the Olacaceae and Icacinaceae, quite clearly on anatomical grounds. For further information see Metcalfe & Chalk, Anatomy of the Dicotyledons 1: 380 (1950).

Inflorescence racemose . . . . . . 1. **Opilia**
Inflorescence umbellate or subumbellate . . . 2. **Rhopalopilia**

1. OPILIA

Roxb., Pl. Corom. 2: 31, t. 158 (1802); Oliv., F.T.A. 1: 352 (1868); Sleumer in E. & P. Pf., ed. 2, 16B: 38 (1935)

Trees, shrubs or woody climbers. Leaves glabrous to pubescent or tomentose; veins often prominent beneath and sometimes above; petiole relatively short. Inflorescences axillary, fasciculate in short or elongate catkin-li racemes when mature; resembling small cones when young, covered by imbricate, peltate, caducous bracts. Flowers small, ♀, solitary or fasciculate, usually in threes or multiples thereof. Petals 4–5, free, usually recurved, caducous; disk-glands free, fleshy, alternating with the stamens. Ovary unilocular; ovule pendulous; style short; stigma truncate. Fruit approximately ellipsoid. Seed large, stone-like.

A palaeotropical genus containing about 25 species, ten of which occur in Africa.

NOTE. Two species described from eastern Africa as *Opilia* are excluded:
O. mildbraedii Engl. in E.J. 54: 291 (1917); Louis & J. Léon. in F.C.B. 1: 283 (1948). Type: Congo Republic, Ruwenzori, Semliki R., *Mildbraed* 2748 (B, holo.!). This specimen is in fact *Thecacoris lucida* Hutch. (Euphorbiaceae).
O. obovata A. Peter, F.D.O.-A. 2, Anhang: 12, t. 17/3. Type: Tanganyika, Kilimanjaro region, *Peter* 17120 (B, holo. †). The illustration shows a species of *Embelia* (Myrsinaceae), which was so identified by Sleumer in E. & P. Pf., ed. 2., 16B, Nachtr.: 339 (1935). However, the specimen was destroyed, making confirmation from authentic material now impossible.

Flowers appearing with mature leaves; branches
always unarmed; peduncle over 3·0 cm. long
when mature . . . . 1. *O. celtidifolia*
Flowers appearing before leaves (or before leaves are
mature); branches often appearing to be armed
with spines; peduncle under 2·0 cm. long when
mature . . . . 2. *O. campestris*

1. O. celtidifolia (*Guill. & Perr.*) *Walp.* in Rep. Bot. Syst. 1: 377 (1842);
P.O.A. C: 168 (1895); Sleumer in E. & P. Pf., ed. 2, 16B: 38, fig. 21/K
(1935); Louis & J. Léon. in F.C.B. 1: 282 (1948); T.T.C.L.: 396 (1949);
F.P.S. 2: 290 (1952); F.W.T.A., ed. 2, 1: 651 (1958); K.T.S.: 351 (1961);
F.F.N.R.: 41 (1962); Garcia in F.Z. 2: 336 (1963). Type: Senegal,
Leprieur (P, holo.!)

Scandent evergreen shrub, sometimes erect, up to 10 m. high, branching
from near the base. Young branches usually green, sometimes reddish,
glabrous to tomentellous; older bark grey to dark brown; lenticels paler,
anastomosing to form shallow longitudinal ridges of cork. Petiole 3–8 mm.
long; leaf-blade lanceolate to ovate or elliptic, 5–12 cm. long, 2–5 cm. wide,
acute or bluntly acuminate to rounded, cuneate to more or less rounded at
the base, coriaceous, more rarely chartaceous, glabrous to tomentellous;
upper surface glossy, often covered with raised dots between tertiary veins
(caused by cystoliths) or slightly tomentellous on both surfaces; secondary
veins 2–7(–8) pairs. Inflorescence axillary, racemose, solitary or clustered
(fasciculate), at first compact (cone-like in appearance), and covered with
peltate imbricate caducous slightly pubescent bracts, ultimately extending up
to 5·5 cm. long (catkin-like); peduncle pubescent; pedicels inserted in groups
or solitary, pubescent with indumentum varying from yellowish-green to
rust-brown. Flowers small, 5-merous, sweet-scented. Petals cream to yellowish-green, oblong-lanceolate, up to 2 mm. long, 0·7 mm. wide. Stamens up
to 2 mm. long, basally attached to petals, sometimes free; filaments filiform.
Fleshy disk-glands exposed by the recurving petals, truncate, under 1 mm.
high. Ovary conical; style truncate. Fruit a drupe, ellipsoid, puberulous to
shortly tomentellous, up to 2·5 cm. long and 1·4 cm. across, yellow to orange
when ripe. Stone large, up to 1·8 mm. long and 1·2 mm. across.

var. celtidifolia

Young branches glabrous to slightly puberulous; leaf-apex tending to be more acute
than rounded.

UGANDA. Acholi District: Aswa R., Abbia Ferry, Mar. 1935, *Eggeling* 1677!; Bunyoro
District: Butiaba Flats, Oct. 1933, *Eggeling* 1442!; Teso District: Serere, Jan. 1933,
Chandler 1059!
KENYA. Northern Frontier Province: Moyale, 22 Aug. 1952, *Gillett* 13745!; Masai
District: NW. of Magadi, 28 Dec. 1958, *Greenway* 9550!; Kwale District: about
19 km. S. of Mombasa, Twiga, 3 Apr. 1963, *Verdcourt* 3609!
TANGANYIKA. Mwanza District: Kalemera, 30 Nov. 1953, *Tanner* 1866!; Tanga District:
Mtimbwani, 6 Dec. 1935, *Greenway* 4221!; Pangani District: Bushiri Estate, 28 Nov.
1950, *Faulkner* 743!
DISTR. U1–3; K1, 4–7; T1–4, 6–8; occurring throughout tropical Africa southwards to
Angola

SYN. *Groutia celtidifolia* Guill. & Perr. in Fl. Senegamb. Tent. 1: 101, t. 22 (1831)
[*Opilia amentacea* sensu Oliv., F.T.A. 1: 352 (1868); Bak. f. in J.L.S. 40: 43
(1911); Thonner, Blutenpflanz. Afr., t. 36 (1908) & Fl. Pl. Afr., t. 36 (1915),
non Roxb.]
O. angiensis De Wild. in Rev. Zool. Afr. 10, Suppl. Bot.: 13 (1922) & Pl. Bequaert.
2: 23 (1923). Type: Congo Republic, Kivu Province, Angi, *Bequaert* 5829
(BR, holo.!)
O. ruwenzoriensis De Wild. in Rev. Zool. Afr. 10, Suppl. Bot.: 12 (1922); Pl.
Bequaert. 2: 24 (1923). Type: Congo Republic, Ruwenzori, *Bequaert* 3840
(BR, holo.!)

O. parviflora A. Peter, F.D.O.-A., 2: 147 (1932), *nomen nudum*. Based on: Tanganyika, Kilimanjaro region, *Volkens* 1964 (B !)

var. **tomentella** (*Oliv.*) *Lucas* in K.B. 21 : 242 (1967). Type: Mozambique, above Zambezi delta, *Kirk* (K, holo. !)

Young branches tomentellous; leaf-apex tending to be more rounded than acute.

Kenya. S. Kavirondo District: Suna, Sept. 1933, *Napier* 5247 !

Tanganyika. Tabora, *C. H. N. Jackson* 12 !; Mbeya District: " Nyika Plateau " (near Sante), Nov. 1899, *Goetze* 1408 !; Iringa District: Lofia [Losio] R., Jan. 1899, *Goetze* 446 !

Distr. K5; T1, 4, 6, 7; Zambia, Rhodesia and Mozambique

Syn. *O. amentacea* Roxb. var. *tomentella* Oliv. in F.T.A. 1: 352 (1868)
      O. tomentella (Oliv.) Engl. in P.O.A. C: 168 (1895) & in E.J. 43: 173, fig. 1/G–J (1909); Sleumer in E. & P. Pf., ed. 2., 16B: 38, fig. 21/G–J (1935); T.T.C.L.: 396 (1949); Garcia in F.Z. 2: 338, t. 68 (1963)

Note. The difference between these two varieties tends to become less distinct as one moves northwards into Tanganyika. Whereas in the Flora Zambesiaca it has been thought possible to separate these two groups at the specific level, I feel that the differences are so slight that they do not merit this treatment. Field studies may well show that even varietal status is not justified.

Due mainly to the large ecological range of this species, the general habit, leaf shape and texture varies considerably. There are indications that a number of ecotypes exist, but insufficient information is to be found on the labels of specimens so far collected to really clarify the situation. Below are cited some specimens which are rather atypical, but at present do not merit formal taxonomic status. Further gatherings and field studies would be most helpful in the clarification of this group. The specimens are characterized by a reddish bark, the indumentum of the pedicel being rusty brown and the leaf-blade tending to be more constantly lanceolate than in the two previous varieties.

Uganda. Busoga District: Mutai, July 1945, *M. Kibuko* J. 20 !; Mengo District: near Entebbe, Kitubulu, Nov. 1938, *Chandler* 2492 !

Tanganyika. Bukoba District: [Bukoba–] Biharamulo road, 13 Nov. 1948, *Ford* 848 !; Biharamulo District: Ruiga Forest Reserve, Dec. 1958, *Procter* 1106 !; Rungwe District: Tukuyu-Ipana, 22 Aug. 1933, *Greenway* 3600 !

Hab. (of species as a whole.) Ranging widely from coastal bushland and riverine forest to upland rain-forest; 0–1850 m.

2. **O. campestris** *Engl.* in E.J. 43: 173 (1909); Sleumer in E. & P. Pf., ed. 2, 16B: fig. 21/D–F (1935); T.T.C.L.: 396 (1949); K.T.S.: 351 (1961). Types: Tanganyika, Pare District, Sadani–Kwagogo, *Engler* 1655 (B, syn. !) & 1660 (B, syn.) & Sengani–Simba, *Engler* 1617 (B, syn. !) & 1625 (B, syn. !) & Gonja–Kiswani, *Engler* 1559a (B, syn.)

Deciduous shrub, 2–5 m. high, with short leafless branches giving the impression that the shrub is armed with spines; lenticels anastomosing on older wood to give corky ridges as in *O. celtidifolia*; bark grey-brown to blackish. Leaves shortly petiolate; leaf-blade elliptic to broadly ovate or suborbicular, 2–5 cm. long, 2·8 cm. wide, broadly acute to rounded, shortly cuneate, puberulous to tomentose when young, usually glabrous, coriaceous when mature. Inflorescence a raceme borne on short younger branches, covered in peltate almost glabrous bracts, ciliate at margin, early caducous; young inflorescence normally appearing before the leaves or before they are mature, cone-like, but shortly pedunculate at maturity; peduncle up to 2 cm. long; pedicels up to 4 mm., inserted singly or in groups of 3; overall appearance of inflorescence more globular and less catkin-like than in *O. celtidifolia*. Flowers small, 5-merous, sweet-scented. Petals cream to pale yellow, glabrous, less caducous than in *O. celtidifolia*. Stamens free, or basally attached to petals. Ovary small, conical, surrounded by glands; stigma truncate. Fruit a drupe, ellipsoid, slightly beaked when young, shortly tomentose, purplish when mature, up to 1 cm. long. Seed stone-like as in *O. celtidifolia*. Fig. 1, p. 4.

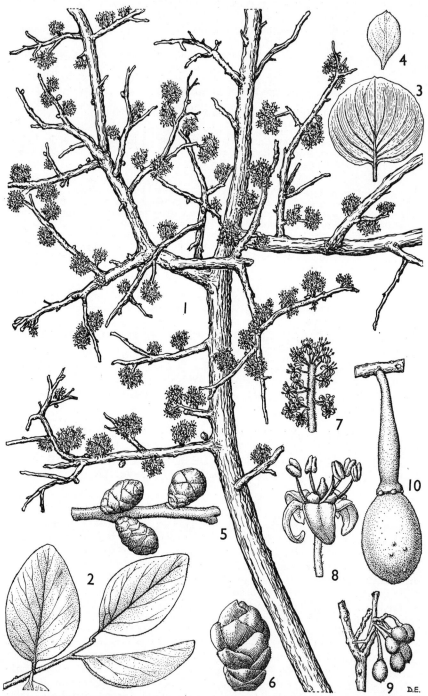

FIG. 1. *OPILIA CAMPESTRIS*—**1,** branch in full flower, × ⅔; **2,** tip of leafy branchlet, × ⅔; **3,** leaf, × ⅔; **4,** same, showing variation, × 2; **5,** part of branchlet with young unexpanded inflorescences, × 2; **6,** young inflorescence, showing overlapping bracts, × 4⅔; **7,** mature inflorescence, × 2; **8,** flower, × 8; **9,** part of branchlet with clusters of fruits, × ⅔; **10,** young fruit, showing remnants of disk-glands, × 3. 1, 6, from *Bally* 8345; 2, from *B. D. Burtt* 3929; 3, 5, from *Semsei* 3407; 4, 9, 10, from *Verdcourt* 3017; 7, 8, from *B. D. Burtt* 818.

KENYA. Northern Frontier Province: Dandu, 30 Mar. 1952, *Gillett* 12657!; Masai District: Namanga–Amboseli road, 20 Nov. 1960, *Verdcourt* 3017! & Nyiri Desert, 10 Oct. 1952, *Bally* 8345!
TANGANYIKA. Shinyanga District: Nindo Forest Reserve, 21 Oct. 1961, *Carmichael* 847!; Dodoma District: Kazikazi, 26 Dec. 1931, *B. D. Burtt* 3425!; Mpwapwa, 8 Feb. 1935, *Hornby* 339!
DISTR. **K**1, 6, 7; **T**1–3, 5; also occurring in Ethiopia, Angola and South West Africa
HAB. Wooded grassland and deciduous bushland, usually associated with *Commiphora*, commonly growing on termite mounds; (70–)850–1450 m.

SYN. *Opilia sp.* sensu Exell & Mendonça, C.F.A.: 331 (1951), quoad *Gossweiler* 9188!, 9197, 10079!

2. RHOPALOPILIA

Pierre in Bull. Soc. Linn. Paris 2: 1236 (1896)

Shrubs or woody climbers. Leaves glabrous to tomentose. Inflorescence axillary, umbellate or subumbellate when mature. Flowers regular and ♀. Petals 4–5, free, reflexed, rapidly caducous. Stamens free or attached basally to petals, caducous. Receptacle cupuliform, ± 4–5-lobed. Disk-glands fleshy. Ovary small, bearing 1 pendent ovule. Fruit subglobose to ellipsoid. Seed large, stone-like.

A genus of about 12 species with but a single representative in East Africa, and four further species in Central and West Africa.

R. umbellulata (*Baill.*) *Engl.* in E.J. 43: 175 (1909); T.T.C.L.: 397 (1949); Garcia in F.Z. 2: 338, t. 69 (1963). Type: Zanzibar I., *Boivin* (P, holo.!)

Scandent shrub or climber; older bark grey to pale brown, with longitudinal striations caused by anastomosing lenticels; younger branches greenish, with buff-coloured narrowly elliptic lenticels having their long axis parallel with the branch-axis; twigs glabrous. Leaf-blade lanceolate to ovate-elliptic, 4–8 cm. long, 1·5–3·5 cm. wide, acute, equal or slightly unequal-sided at base, cuneate, coriaceous, glabrous. Inflorescences axillary pedunculate bracteate umbels; peduncle up to 2 cm. long, ± pubescent, with pedicels attached to a swollen cushion-like expansion of the peduncle; pedicels slender, up to 5 mm. long; bracts very caducous. Flowers small, 5-merous; petals ± 2 mm. long, 0·5 mm. wide, glabrous inside, puberulous outside, recurved, very caducous as are the stamens which are usually basally attached to the petals. Stamens ± 1·5 mm. long. Glands and ovary exposed by loss of other floral parts; ovary elongated, conical; stigma truncate. Fruit a drupe, subglobose, up to 1 cm. in diameter, yellowish-orange when ripe. Fig. 2, p. 6.

KENYA. Kilifi District: Mida, Feb. 1929, *R. M. Graham* in *F.D.* 1830! & Arabuko, Mar. 1930, *R. M. Graham* in *F.D.* 2301!
TANGANYIKA. Tanga District: Sawa, 29 Sept. 1956, *Faulkner* 1919!; Pangani District: Msumbugwe Forest Reserve, Sept. 1955, *Semsei* 2250!; Uzaramo District: Kisiju, Sept. 1953, *Semsei* 1373!
ZANZIBAR. Zanzibar I., Kombeni caves, July 1930, *Vaughan* 1424! & Chukwani, 28 July 1930, *Vaughan* 1433! & near Chukwani, 2 Nov. 1959, *Faulkner* 2390!
DISTR. **K**7; **T**3, 6; **Z**; confined to the coastal strip and Zanzibar as far south as Mozambique
HAB. Coastal forest (dry evergreen and ? edges of lowland rain-forest), *Brachystegia* woodland and coastal bushland; 0–100 m.

SYN. *Opilia umbellulata* Baill. in Adansonia 8: 199 (1868)
O. sadebeckii Engl. in N.B.G.B. 2: 282 (1899). Types: Tanganyika, Pangani, *Stuhlmann* 659 & Zanzibar I., *Stuhlmann* 661 (both B, syn.!)

FIG. 2. *RHOPALOPILIA UMBELLULATA*—**1,** flowering branches, × ⅔; **2,** inflorescence, × 3; **3,** flower, × 6; **4,** flower with perianth and stamens removed to show disk-glands and pistil, × 8; **5,** part of fruiting branch, × ⅔. 1, 2, from *Paulo* 127; 3, 4, from *Faulkner* 681; 5, from *Faulkner* 2390.

INDEX TO OPILIACEAE